AI Made Simple

No Tech Degree Required: Plain Language on How Artificial Intelligence Works and How to Use it Today.

Dr. Rosie Milligan

Printed and Bound in the United States of America

Published And Distributed By
Professional Publishing House, LLC
Los Angeles, California
Telephone: 323-750-3592
Email:professionalpublishinghouse@yahoo.com
www.Professionalpublishinghouse.com

Cover design: TWA Solutions
First printing March 2026
979-8-994-1309-1-9
10987654321

For inquiry, contact Drrosie@aol.com.

Dedication

I thank God for granting me the strength, wisdom, and compelling desire—at the age of 80—to continue sounding urgent warnings to the people. I do not take this assignment lightly. It is purpose-driven, and it is necessary.

My legacy to all people is this:

I leave you love.

I leave you hope.

I leave you a hunger for knowledge—for you and for your family.

I dedicate this book to my family and friends, near and far. I pray that you catch a glimpse of what is coming in this world, and more importantly, that you prepare for the new technology that will change society in a way no other technology ever has.

Artificial Intelligence will reshape how we live, work, earn, and survive. It will create opportunity for some and devastation for others. For those who refuse to grasp its coming, the impact will be nothing short of an economic tsunami.

Table of Contents

A Note from the Author

Artificial Intelligence is not coming. It's already here. And while AI may not replace you, someone who understands AI just might.

I am 80 years old, and I work with Artificial Intelligence and technology the same way a farmer works his mule: every day, steadily, purposefully, and without fear. I use AI for both my personal and business life because I refuse to be left behind by progress. Age is not a barrier, ignorance is.

If you've been lagging behind because technology feels intimidating, let me say this clearly: you do not have to be a techie to use AI. Artificial Intelligence is simple, user-friendly, and accessible. If you can ask a question, you can use AI. And once you do, you'll wonder how you ever lived without it.

AI can save you time. AI can save you money. AI can put information at your fingertips in seconds. It can help you assist your children and grandchildren with math, English, and writing. It can help you plan meals, organize your life, and better understand medical diagnoses, medications, and possible side effects. AI can help you ask better questions, make more informed decisions, and move through life with greater confidence.

AI is not to be feared. It is to be embraced.

That said, Artificial Intelligence is a tool, not a substitute for wisdom. It is not your doctor, and it is not your lawyer. While AI is remarkably helpful and often accurate, it is still a machine. Responsible use matters. Misuse can cost you legally, financially, and ethically. That is why learning how to use AI wisely, and teaching our children to do the same, is not optional. It is essential.

The purpose of this book is simple and intentional: to give you the ABCs of Artificial Intelligence, to remove the fear, break down the confusion, and show you that AI is not complicated, mysterious, or reserved for the young. It is a tool for everyone willing to learn.

If I can do this at the age of 80, so can you.

Let's get started!

About the Author

Dr. Rosie Milligan, CEO of Professional Business Management/Consulting Services and My Tech Academy. She is an author, publisher, and one of Los Angeles's Renowned Financial Gurus, a seasoned Senior Estate Planner, Business Coach, and Business Consultant. She also holds a Ph.D. in Business Administration. She is sought after nationwide for her business and financial coaching services.

She has always been an achiever. Every career or business she's been involved in has included helping other people accomplish what they wanted in life. Her motto, "Erase 'NO,' Step Over 'CAN'T,' and Move Forward With LIFE," has been a motivating influence for hundreds to whom she has been a mentor and role model.

Dr. Milligan's reputation extends nationwide, with individuals and organizations seeking her out for top-tier business and financial coaching services. She has authored thirty-six books and published 450 books for other authors. Her books: *What You Need To Know Before You Start A Business, Departing This Life Preparations, Getting out of Debt Made Simple, Understanding Credit Made Simple, and Creating a Budget Made Simple* have helped many across the country. She is the largest and fastest-growing African American female publisher in the nation.

She is also the host of a weekly live Zoom talk show, *EXPRESS YOURSELF HOUR*, and she is the founder and director of Black Writers on Tour. Contact her at: 323-750-3592; drrosie@aol.com; Drrosie.com.

Introduction

We are living in one of the most transformative moments in human history. Artificial Intelligence—once the stuff of science fiction—is now woven into the very fabric of our daily lives. From the moment you wake up and ask your phone about the weather, to the moment a streaming service recommends your next favorite show, AI is already your quiet, invisible companion.

This book was written for everyone: the grandmother who wonders why her phone seems to 'know' what she needs, the middle schooler fascinated by chatbots, the small business owner curious about cutting costs with technology, and the professional looking to stay ahead of the curve. No technical background is required.

Throughout each chapter, you will find live examples—real prompts and real AI responses—so you can see exactly what AI looks like in practice, not just in theory. You are encouraged to try these prompts yourself using any of the free AI tools mentioned in this book.

Welcome to the age of Artificial Intelligence. Let's explore it together.

Chapter 1
Why AI Matters to Seniors, Too

AI Knowledge, Legacy, and Generational Wealth

This moment matters, and history has taught us why. Those who fail to adapt are left behind, and their children pay the price. Artificial Intelligence is not a trend. It is not a fad. It is not optional.

AI is already shaping how money is made, how information is shared, how children learn, how businesses survive, and how families plan for the future. The question is not *if* AI will impact your life—the question is whether you will be prepared or blindsided.

For too long, seniors have been quietly pushed aside by technology, made to feel that learning something new is "too late," "too hard," or "not for them." That lie has cost families dearly. When seniors are excluded from technology, they lose control over finances, information, healthcare decisions, and even their own legacy. When seniors lose control, generational wealth is at risk.

AI changes that.

AI Gives Seniors Real Power

Artificial Intelligence puts information back into your hands. It allows you to understand what is happening with your money, your property, your business, your health, and your legal affairs, without depending on someone else to explain it to you, or worse, take advantage of you. AI is a tool for protection. AI is a tool for clarity. AI is a tool for independence.

In practical terms, this means real change in daily life. AI-powered devices can remind you to take your medication, help you navigate complex health information, alert family members in an emergency, and even detect falls. Smart speakers respond to your voice—no typing required. AI-enhanced video calls add real-time captions, making it easier to stay connected with family near and far. Virtual assistants can read out your bank statements, explain a legal document in plain English, or walk you through a Medicare decision, so you are never again at the mercy of someone else's interpretation.

This is not just convenience. This is power.

Legacy Is More Than Property—It Is Knowledge

For families who care about leaving something behind beyond just memories, AI matters deeply. Wealth today is not just about property and money. It is about knowledge. The families who pass down knowledge pass down power. The families who do not are forced to start over in every generation.

That cycle must end.

Seniors are the gatekeepers of wisdom, history, and assets. When seniors understand AI, they become stronger decision-makers, better teachers to their children and grandchildren, and wiser stewards of the wealth they worked a lifetime to build. When seniors rise, families rise. When families rise, communities rise. And when communities rise, legacy lives on.

Learning AI today means:

- You can protect your assets instead of losing them through confusion
- You can guide your family instead of being guided by them

- You can participate in the modern economy instead of being shut out
- You can leave instructions, not chaos, behind

Legacy is not just what you leave—it is how well you prepare the next generation to receive it.

You Do Not Have to Be a Tech Expert

You do not need a college degree or a high school diploma. You do not need to understand computers or code. All you need is the willingness to learn and the belief that you can do it. This is not about becoming an expert. This is about becoming empowered.

If you can read, if you can ask questions, if you care about your future and your family's future, you can learn AI. AI is simple, user-friendly, and accessible. If you can ask a question, you can use AI.

And waiting is no longer safe. Waiting means falling further behind. Waiting means depending on others. Waiting means risking everything you worked for. Learning puts you in front of the economic storm—and changes everything.

This Is Why It Matters Now

Artificial Intelligence will reshape how we live, work, earn, and survive. It will create opportunity for some and devastation for others. For those who refuse to engage, the impact will be nothing short of an economic tsunami.

But for seniors who choose to learn—who refuse to be left behind—AI is not a threat. It is a tool. It is protection. It is power. It is the key to a legacy that does not just survive the future but helps shape it.

SAMPLE AI PROMPT

Sample Prompt You Can Try

I am 72 years old and not very comfortable with technology. Can you recommend three simple AI tools that could make my daily life easier and safer? Please explain each one in plain, simple language—no jargon.

Sample AI Response

Of course! Here are three AI tools that are genuinely easy to use and could make a real difference in your day-to-day life:

1. Amazon Echo (Alexa)

Think of this as a friendly helper you just talk to—no buttons, no screens required. You can ask it to set a reminder for your medication, play your favorite music, call a family member, or even tell you the weather. It sits on your counter and listens for your voice. Many seniors find it life changing.

2. ChatGPT (at chat.openai.com)

This is like having a very patient, knowledgeable friend available 24 hours a day. You can type a question—just like sending a text—and it will answer in plain English. Ask it to explain a letter from your doctor, help you write a birthday message, or suggest a recipe. It never gets frustrated and never rushes you.

3. Google Photos (on your smartphone)

If you have a smartphone, this app automatically organizes all your photos by people, places, and dates—with no effort from you. It can find a photo of your grandchildren from three years ago in seconds. It also backs up your photos, so they are never lost if your phone breaks.

Start with just one of these. Take your time. There is no rush, and you cannot break anything by trying.

Try this prompt at chat.openai.com or claude.ai—type it in just as written above.

Chapter 2
Why AI Matters to Children and Older Students

Our Children Cannot Afford to Be Left Behind

AI is not just for adults, businesses, or tech people. AI matters deeply for our children and older students, and the sooner parents understand that, the better off our families and communities will be.

Many parents are afraid of AI. They worry it will make their children lazy, kill their creativity, or stop them from thinking for themselves. That fear is understandable, but it is misplaced. AI is not here to replace thinking. It is here to support learning. Just like calculators did not destroy math and books did not destroy memory, AI will not destroy our children's minds. The real danger is not AI. The real danger is leaving our children unprepared for the world.

Children Learn Early—and Fast

Research has shown us something powerful: by the age of six, children have already developed a significant portion of the habits and mindsets that will shape their entire lives—how they think, how they learn, how confident they feel about knowledge. That is why I firmly believe AI should be introduced as early as kindergarten.

Children today are growing up in a digital world, whether we like it or not. Tablets, phones, smart TVs, and computers are already part of their environment. AI is simply the next layer of that world, and when introduced correctly, it becomes a tool, not a crutch. Used properly, AI can:

- Help children ask better questions,
- Strengthen reading and comprehension,
- Support math learning step by step, and
- Encourage curiosity instead of frustration.

AI does not tell children what to think. It helps them learn how to think.

Parents Are Overwhelmed—And That Is Real

Today's parents are busy. Many are working one job, two jobs, or juggling life just to survive. When they get home, they are tired. They love their children, but they do not always have the time, energy, or training to help with schoolwork the way they want to. On top of that, school is taught differently now. Math is not taught the way many of us learned it. Parents look at homework and say, "This is not how I learned it." And they are right. That leaves parents feeling inadequate, frustrated, and sometimes embarrassed.

AI helps fill that gap. AI can explain math problems step by step using today's methods. It can break down reading assignments in plain language. It can help a child understand a lesson without making them feel stupid. That is not replacing parenting. That is supporting parenting.

When Resources Are Missing, AI Becomes the Bridge

Many inner-city schools do not have enough books. That is a fact. Many families struggle with literacy. That is a fact. Libraries are closing, underfunded, or unsafe for children to attend alone. That is a fact.

So what happens when children need help, but the traditional support systems are gone? AI becomes the answer.

With proper guidance, AI can:

- Read stories aloud to children
- Help struggling readers practice safely at home
- Support homework when no tutor is available
- Give children access to information their schools may lack

AI does not judge. AI does not shame. AI is available 24 hours a day. For families who have been locked out of resources for generations, AI becomes an equalizer.

Older Students Need AI, Too

AI is not just for little children. Middle school students, high school students, college students, and adult learners all benefit—and especially those who fell behind or were never properly supported. Many older students are not lazy. They are discouraged. AI helps them re-enter learning without fear.

The World Economic Forum estimates that many of the jobs that will exist in twenty years have not been invented yet. What is certain is that AI fluency—the ability to understand, work with, and think critically about AI—will be one of the most valuable skills a young person can have. Learning about AI now is not just interesting. It is career preparation. Fields ranging from medicine and law to music and agriculture are being transformed. Understanding the technology early gives young people a significant advantage.

Older students use AI to:

- Organize essays and sharpen their thinking,
- Study for exams across any subject,
- Understand difficult concepts at their own pace, and
- Build confidence in learning again.

Creativity Is Not Lost—It Is Expanded

AI does not destroy creativity. Lack of opportunity destroys creativity. When children have the support they need to understand, they become confident. When they are confident, they create. When they create, they innovate.

AI can help a child write stories, explore ideas, ask "what if?" questions, and build projects they never thought possible. Far from limiting imagination, AI can be a powerful creative partner—as long as students understand its role as a tool, not a replacement for their own thinking and voice.

Creativity does not disappear when children are supported. It blossoms.

The Parent's Role Is Unchanged

AI does not replace parents. AI does not replace teachers. AI does not replace guidance, discipline, or values. Parents still teach right from wrong, responsibility, balance, and critical thinking. AI is a learning assistant, not a parent, not a teacher, not a moral compass.

Parents who embrace AI are not giving up control. They are giving their children a fighting chance.

Fear Will Not Protect Our Children—Education Will

We cannot protect our children by hiding technology from them. We protect them by teaching them how to use it wisely. AI is already here. The question is not whether our children will encounter it. The question is whether they will be prepared to use it well—or left behind, or worse, led to use it in ways that harm them.

Parents who embrace AI are not reckless. They are forward-thinking. They are protecting their children's future. And in this fast-moving world, preparing your children is one of the most powerful acts of love a parent can offer.

SAMPLE AI PROMPT

Sample Prompt You Can Try: AI as a Student Tutor

I am a 14-year-old student struggling to understand photosynthesis for my biology class. Can you explain it in a fun, easy-to-understand way? Use an analogy that a teenager would find interesting.

Sample AI Response

Great question! Let me make photosynthesis click for you.

Imagine your phone battery is dead. You need to plug it in to charge it up so it can do all the cool stuff—text, play games, stream videos. Plants have the same problem: they need energy to grow, make flowers, and stay alive.

Here is the wild part: plants charge themselves using SUNLIGHT. Like tiny solar panels built right into their leaves.

Here is how it works: The leaf sucks in water from the roots (like drinking through a straw) and pulls in carbon dioxide gas from the air (the stuff we breathe OUT). Then—and this is the cool part—the green stuff in the leaf (called chlorophyll) uses sunlight to smash those ingredients together and make sugar. That sugar is basically the plant's food and energy source.

Oh, and the leftover "waste product"? That is pure oxygen—the stuff we breathe in. So, plants are literally cleaning our air while feeding themselves. Pretty genius engineering, honestly.

Quick formula to remember: Water + Carbon Dioxide + Sunlight = Sugar + Oxygen.

You are basically breathing right now because a plant was hungry. Mind blown? It should be. □ *Notice how AI adapts its tone and examples to the age and context described in the prompt.*

Chapter 3
History of AI

AI: How We Got Here

Artificial Intelligence did not just appear overnight. It has been around for decades—quietly evolving, steadily growing, and waiting for the moment when everyday people could finally access its power. It did not start with smartphones or the internet. It started with a question.

The 1950s: The First Question

In 1950, a quiet British mathematician named Alan Turing asked something that would change the world: *"Can machines think?"*

He did not mean think like humans—with feelings, dreams, or consciousness. He meant think well enough to solve problems, make decisions, and learn patterns.

He even created a test—now known as the Turing Test—to determine whether a machine could converse so convincingly that a human could not tell it apart from another person. That single question opened the door to everything that followed.

Six years later, in 1956, a group of visionary researchers gathered at a college workshop and officially gave this idea a name: Artificial Intelligence. Their goal was straightforward—teach computers to do things humans do, like solve problems, understand language, and make decisions. In those early years, AI was used mainly by universities and the military to solve math problems, decode messages, and play strategy games like chess.

The dream was bold. The optimism was real. And the hard work had just begun.

The 1960s–1970s: Big Hopes, Slow Progress

People believed AI would grow fast. It did not. Computers were too slow, too expensive, and did not have enough information to work with. AI at this stage could only do exactly what it was told—nothing more, nothing less. Researchers hit walls they had not anticipated, funding dried up, and the field entered what became known as the "AI Winter"—a period of reduced enthusiasm and investment.

But the idea never died. It simply waited for better tools.

The 1980s: AI Becomes a Helper

In the 1980s, AI found its footing through something called expert systems—programs designed to copy how human experts think and make decisions. Doctors used them to help diagnose illnesses. Engineers used them to detect equipment problems. Businesses used them to guide complex decisions.

These were real, practical applications—and they worked. But they still had a critical limitation: they could not learn on their own. Every rule had to be programmed by hand. They were powerful assistants, but they were not yet truly intelligent.

The 1990s–2000s: Machines Begin to Learn

This is where everything changed.

Instead of programming every instruction manually, researchers discovered something remarkable: computers could be taught to learn from data. The more information they received, the better they got—on their own. This breakthrough became what we now call machine learning, and it quietly transformed the world.

In 1997, IBM's Deep Blue defeated world chess champion Garry Kasparov—a landmark moment that proved machines could master even the most complex human games. AI began powering credit scoring, fraud detection, search engines, and speech recognition. It was working behind the scenes of daily life long before most people realized it.

The 2010s: AI Moves Into Everyday Life

With faster computers, widespread smartphones, and the explosion of the internet, AI moved out of laboratories and into living rooms. The 2010s brought facial recognition, real-time translation, voice assistants like Siri and Alexa, and recommendation engines that seemed to know what you wanted before you asked.

In 2022, the release of tools like ChatGPT brought AI directly into the hands of everyday people—not just scientists, engineers, or large corporations, but seniors, parents, students, small business owners, and families building a legacy. For the first time in history, the power of AI was truly accessible to all.

Why This History Matters to You

AI was never meant to replace people. It was designed to assist people. What changed is access.

What once belonged only to elite institutions and government agencies is now available to anyone willing to learn. Today, AI helps people write and edit, translate languages, get directions, plan finances, understand medical information, support their children's education, and build businesses from the ground up.

AI did not just arrive. It evolved through decades of questions, setbacks, breakthroughs, and determination. It was built by curious minds who refused to accept that machines could not learn.

And now, it is your turn. The technology is here. The access is here. The only question left is whether you are ready to use it wisely.

SAMPLE AI PROMPT

Sample Prompt You Can Try—AI Tells Its Own History

Give me a brief, engaging timeline of the most important milestones in AI history, from the 1950s to today. Format it as a story, not a list. Make it feel like a dramatic journey.

Sample AI Response

It began with a question.

In 1950, a quiet British mathematician named Alan Turing asked something that would change the world: "Can machines think?" He proposed a test—if a machine could converse so convincingly that a human couldn't tell it apart from another human, perhaps it deserved to be called intelligent. The world wasn't ready for the answer.

Six years later, in a summer conference at Dartmouth College, a group of visionary scientists gave the dream a name: Artificial Intelligence. They were boldly optimistic. They believed machines would match human intelligence within a generation.

They were wrong—but magnificently so.

Decades of progress and painful setbacks followed. Programs that could play chess. Programs that could prove theorems. And then, silence—the "AI Winters," when funding dried up and the dream seemed to stall.

Then came the 1990s and a new idea: instead of programming intelligence, what if machines could learn? In 1997, IBM's Deep Blue defeated world chess champion Garry Kasparov. The machine had studied millions of games. It had learned.

By the 2010s, something extraordinary happened. With mountains of data, powerful computers, and layered "neural networks" inspired by the human brain, AI suddenly began to see, to hear, and to speak. It recognized faces. It translated languages in real time.

And then, in late 2022, it spoke to all of us through tools like ChatGPT, available to anyone with an internet connection.

Seventy years after Turing's question, the machines had arrived. Not to think as we think—but to help us think better.

The same facts can be delivered as a dry list or as a compelling story—the prompt makes all the difference.

Chapter 4
Definitions for AI

Plain Language First—Always

Before we go any further, let us make sure we are speaking the same language. One of the biggest reasons people feel intimidated by AI is the vocabulary. The terms sound complicated, technical, and foreign. But once you understand what they actually mean— in plain English—the fear disappears.

This chapter breaks down the most important AI terms simply and clearly. No jargon. No engineering degree required.

What Is Artificial Intelligence (AI)?

Artificial Intelligence is simply a computer that has been taught how to think a little like a human. It can learn, remember, answer questions, solve problems, and help you get things done faster—but it only knows what it has been taught. AI does not have a mind of its own. It does not have feelings, opinions, or a personal agenda. It works for you.

Think of AI like a very smart assistant that never gets tired, never gets frustrated, and is available every hour of every day.

What Is Chat?

Chat simply means you are having a conversation with the computer—typing questions or speaking—and it responds. Instead of clicking through buttons and menus trying to find what you need, you just ask. Like you would ask a person.

No special commands. No confusing instructions. Just a conversation.

Chat is simply talking back and forth.

What Is GPT?

GPT stands for Generative Pre-trained Transformer. Do not let that intimidate you—the technical name does not matter. What matters is what it does.

In plain English: GPT is the brain behind the chat. It is the part of the system that has read millions of books, articles, websites, and conversations, so it knows how to put words together in a way that makes sense. GPT is why the computer can write, explain, summarize, and speak in everyday language rather than cold computer code.

The "generative" part means it can create—new sentences, new ideas, new explanations—not just repeat memorized answers.

Putting It All Together—AI + Chat + GPT

Here is how the three pieces work together:

- **AI** is the helper—the intelligent system doing the work,
- **Chat** is the conversation—the way you communicate with it,
- **GPT** is the brain—what makes the conversation smart, natural, and useful,

Bottom line: Artificial Intelligence is a smart tool you can talk to, ask questions, write with, learn from, and use to make life and business easier—without being a tech expert.

Other Terms Worth Knowing

As you use AI, you will encounter a few other words that are helpful to understand:

Machine Learning—Teaching a computer to improve by giving it lots of examples, the same way a child learns to recognize a dog, not from a rulebook but from seeing many dogs. The more it sees, the better it gets.

Prompt—The question or instruction you type into AI. Think of it like placing an order at a restaurant. The more specific your order, the better your result. A vague prompt gets a vague answer. A clear, detailed prompt gets a clear, useful answer.

Large Language Model (LLM)—A fancy name for AI systems like ChatGPT and Claude that have been trained on enormous amounts of text. They understand language, answer questions, write content, and hold conversations. The "large" simply means they learned from a very large amount of information.

Hallucination—This one is critical to understand. Sometimes, AI will state something incorrectly with complete confidence. It is not lying—it simply generates what sounds plausible based on patterns, without always checking whether it is actually true. This is why you should always verify important information—especially anything related to health, finances, or legal matters—with a qualified professional.

Algorithm—A set of step-by-step instructions a computer follows to complete a task or make a decision. When your bank flags a suspicious charge, an algorithm made that call. When Netflix recommends a show, an algorithm chose it.

The Most Important Thing to Remember

AI is a tool. You are the one in charge.

AI does not make decisions for you. It does not run your life. It does not replace your wisdom, your judgment, or your experience. It answers when you ask. It helps when you direct it. And it only goes where you point it.

Understanding these definitions is not about becoming technical. It is about becoming confident. Once you know what these words mean, you will never feel talked over, confused, or left out of a conversation about AI again.

Knowledge is power. And now you have it.

SAMPLE AI PROMPT

Sample Prompt You Can Try—AI Terms in Plain English

Explain these 5 AI terms as if you're talking to someone who has never studied technology: Machine Learning, Neural Network, Prompt, Large Language Model, and Hallucination. Use a simple real-world analogy for each one.

Sample AI Response

Great questions! Here is each term explained with an everyday comparison:

Machine Learning: Think of teaching a child to recognize a dog. You don't write out a rulebook saying "four legs, fur, tail." You just show them hundreds of dogs and say "dog!" until they get it. Machine learning works the same way—instead of rules, we show the computer millions of examples and let it figure out the pattern on its own.

Neural Network: Imagine a factory assembly line where each worker passes a product to the next, each one doing a small job that builds on the last. A neural network is like that—thousands of tiny computing "workers" arranged in layers, each one processing a tiny piece of information and passing it along, until the final result comes out the other end.

Prompt: A prompt is simply the question or instruction you give to an AI. It is like placing an order at a restaurant. The more specific your order ("a medium coffee, oat milk, no sugar"), the better your result. A vague order ("something warm") gets you something unpredictable.

Large Language Model: Imagine someone who has read virtually every book, article, and website ever written—and can now have a conversation drawing on all of that reading. That is essentially what a large language model is. It has processed billions of words and learned the patterns of human language.

Hallucination: This one is important! Sometimes an AI will state something false with complete confidence—like a student who did not study but still raises their hand. The AI is not lying; it simply generates what sounds plausible based on patterns, without checking whether it is actually true. Always verify important facts.

You can ask AI to explain any technical concept using this format: 'Explain [topic] using a simple analogy for a complete beginner.'

Chapter 5
How You Are Already Using AI

Here is a fact that might surprise you: you have been using AI for years. Every time you use face recognition to unlock your phone, ask Siri a question, get autocorrect suggestions, or have photos automatically organized, you are using AI.

Netflix, Spotify, and YouTube use AI to recommend content. Google Maps uses AI to route around traffic. Your bank uses AI to detect fraud. Hospitals use AI to read X-rays. AI is already woven into daily life.

SAMPLE AI PROMPT

Sample Prompt You Can Try—A Day in the Life with AI

Walk me through a typical day for a retired person living alone—from waking up to going to bed—and highlight every moment where AI is silently at work behind the scenes. Make it feel like a revelation.

Sample AI Response

Your alarm goes off at 7:15 a.m.—but you did not set that time randomly. Your phone studied your sleep patterns over weeks and gently nudged you to wake during your lightest sleep phase. That was AI.

You ask your smart speaker what the weather looks like today. It checks seventeen different data sources, weighs local sensor readings against satellite data, and gives you an answer in two seconds. AI again.

You make coffee and open your email. Overnight, forty-three messages arrived. You only see twelve—the rest were quietly sorted into spam or promotions by an AI that has been learning your preferences for years.

You call your bank about a charge. Before you even speak to anyone, an AI has already scanned your account, noticed that the charge is from a merchant you have never used in a city you have never visited, and flagged it for review.

After lunch, you watch a little television. The three shows your streaming service recommended? Each one was chosen by an algorithm that analyzed not just what you have watched, but how long you watched it, at what time of day, and what similar viewers enjoyed after finishing the same series you just completed.

You go for a walk and check your route on your phone. The map knows there is a slow-down three blocks ahead because thousands of other phones moving slowly through that intersection told it so.

By the time you sit down for dinner, AI has touched nearly every convenience of your day. Not intrusively. Not dramatically. Just quietly, invisibly, making things work a little better.

That is the world we already live in.

AI already touches nearly every convenience of daily life—most people are simply unaware of it.

Chapter 6
The Dos and Don'ts When Using AI

Simple Rules That Protect You and Maximize Your Results

AI is one of the most powerful tools available to everyday people today—but like any powerful tool, it works best when you know how to use it correctly. A hammer builds a house in the right hands and causes damage in the wrong ones. AI is no different.

These dos and don'ts are not complicated. They are common sense guidelines that will help you get the most out of AI while protecting yourself, your family, and your finances.

THE DOS

1. Do Provide a Clear Prompt

Prompting is simply telling AI exactly what you want. You type your question, request, or instruction into the chat box—and AI responds. The clearer and more specific your prompt, the better your result.

Examples of prompts you can use right now:

- *"Explain the side effects of metformin in plain language."*
- *"Create a weekly meal plan for a diabetic senior on a budget."*
- *"Help me write a letter to my landlord about a repair that has not been made."*
- *"Solve this math problem step by step: ..."*
- *"Create an outline for a small business plan for a home cleaning service."*

Think of a prompt like placing a specific order at a restaurant. The more detail you give, the better your meal turns out.

2. Do Iterate—Try Again If the First Answer Is Not Right

If AI gives you a response that is not quite what you were looking for, do not give up. Adjust your prompt and try again. Add more detail. Rephrase the question. Ask it to simplify, expand, or approach the topic differently. This is called iterating, and it is one of the most important skills in using AI well. The first answer is not always the best answer—and you are always allowed to ask again.

3. Do Fact-Check Important Information

Remember: AI is a machine, not a human. It does not always get things right. When AI gives you important information—especially about health, money, legal matters, or current events—verify it. Ask AI where the information comes from. Look it up from a trusted source. Cross-check with a professional when the stakes are high.

A good habit: treat AI responses the way you would treat advice from a very knowledgeable friend. Helpful and worth hearing—but always worth confirming before you act on it.

4. Do Use AI as a Starting Point, Not an Endpoint

AI is excellent at giving you a strong first draft, a starting outline, a list of options, or a general explanation. But your own judgment, your own experience, and your own wisdom always have the final word. Use AI to open the door—then walk through it yourself.

THE DON'TS

1. Don't Rely on AI Completely

AI is a tool, not a crutch. Do not hand your thinking over to a machine. Use AI to support your decisions—not to make them for you. Your discernment, your values, and your life experience are irreplaceable. No algorithm can substitute for wisdom earned through living.

2. Don't Use AI Maliciously

AI must never be used to harm, deceive, manipulate, or take advantage of others. That includes generating false information about people, creating scam messages, spreading misinformation, or using AI to cheat or defraud. Misuse of AI carries real consequences—legally, financially, and ethically. Use it to build, not to destroy.

3. Don't Use AI as Your Doctor

AI can be helpful for understanding health information—looking up the side effects of a medication, learning more about a diagnosis, exploring dietary information related to a specific condition, or preparing questions before a doctor's appointment. That kind of use is appropriate and smart.

What AI must never replace is actual medical care. Do not use AI to diagnose yourself. Do not use AI to decide on a treatment plan. Do not let an AI response talk you out of—or into—a medical procedure. Your doctor has your full history, your test results, and professional training. AI does not. When your health is on the line, see a professional.

4. Don't Use AI as Your Lawyer

AI can help you understand general legal concepts, learn what certain terms mean, or get a broad sense of your rights in a situation. But AI is never a substitute for an attorney. Laws vary by state. Details matter enormously in legal situations. A mistake in a legal matter can cost you your home, your savings, your business, or your freedom. Never let an AI response replace a consultation with a licensed attorney when the situation calls for one.

5. Don't Share Sensitive Personal Information

Never enter your Social Security number, bank account details, passwords, or private medical records into an AI chat. Treat AI conversations the way you would treat a semi-public space—helpful and accessible, but not the place for your most private information.

The Bottom Line

AI used wisely is a remarkable gift. It saves time, solves problems, fills knowledge gaps, and levels the playing field for people who have historically been locked out of information and resources.

AI used carelessly can cost you.

Know the rules. Follow them. And AI will serve you well.

SAMPLE AI PROMPT

COMPARISON: Vague vs. Specific Prompts

VAGUE PROMPT
Tell me about health.

AI Response:

Health is a broad and important topic that covers many areas of life. Good health generally includes physical health, mental health, emotional well-being, and social connection.

Physical health involves regular exercise, a balanced diet, adequate sleep, and routine medical checkups. Mental health is equally important and includes managing stress, maintaining positive relationships, and seeking help when needed. Preventive care—such as vaccinations, cancer screenings, and regular blood pressure checks—plays a key role in staying healthy throughout life.

If you have specific health concerns, please consult a qualified healthcare professional for personalized guidance.

SPECIFIC PROMPT

I am a 68-year-old woman with Type 2 diabetes. Can you explain in simple terms what foods I should prioritize and which I should limit to better manage my blood sugar? Please give me 3 practical, easy-to-follow tips.

AI Response:

Managing blood sugar through food choices is one of the most powerful tools you have—and small changes can make a real difference. Here are three practical tips:

1. Choose slow-digesting carbohydrates. Foods that release sugar slowly into your bloodstream are your best friends. Think: whole grain bread instead of white bread, brown rice instead of white rice, oatmeal instead of sugary cereals, and sweet potatoes instead of regular potatoes. These keep your blood sugar steady rather than causing sharp spikes.

2. Fill half your plate with non-starchy vegetables. Vegetables like broccoli, spinach, green beans, cucumbers, and cauliflower have very little effect on blood sugar and are packed with nutrients. Making

them the centerpiece of your meals naturally reduces the room for foods that spike sugar.

3. Limit sugary drinks and processed sweets. Regular soda, fruit juice, sweetened coffee drinks, cookies, and pastries cause some of the fastest blood sugar spikes. Water, sparkling water, or unsweetened herbal tea are much better choices for hydration.

One bonus tip: eating in a consistent pattern—meals at regular times rather than skipping and then overeating—also helps your body regulate blood sugar more effectively.

Please discuss any significant dietary changes with your doctor or a registered dietitian, as they can tailor advice specifically to your health situation and medications.

Same topic. Completely different results. Specificity is the single most powerful tool in using AI well.

Chapter 7
AI Tools—What They Are and How to Choose Wisely

There Are Thousands of Tools—You Only Need a Few

AI tools include anything that uses artificial intelligence to help you get something done. The landscape has exploded in recent years—there are now thousands of AI tools available, and more are being created every day. That can feel overwhelming. But here is the truth: you do not need most of them. You need the ones that solve your specific problems, and you need to learn those well.

Do not chase every new tool you hear people talking about. Find what works for you, master it, and let it work for you consistently. Depth beats distraction every time.

The Main Categories of AI Tools

AI tools generally fall into these categories:

- Chatbots and Conversational Assistants
- Image, Video, and Audio Generation
- Productivity and Writing Assitance
- Code Generators and Developer Tools
- Business Automation and Analysis

Chatbots and Conversational Assistants

These are the tools you talk to—the ones you type questions into and receive answers from in plain language. They are the most accessible entry point into AI for most people.

The three most widely used are:

- **ChatGPT** (by OpenAI)—the most well-known, excellent for writing, answering questions, planning, and research
- **Gemini** (by Google)—deeply connected to Google's ecosystem; helpful if you use Gmail, Google Docs, or Google Search
- **Claude** (by Anthropic)—known for thoughtful, detailed responses and strong performance on longer documents and nuanced topics

All three have free versions. All three are powerful. Start with one, learn it well, and expand from there.

Image, Video, and Audio Generation

These tools create visual and audio content from a simple text description. You type what you want—and AI builds it.

- Image tools like Midjourney and DALL-E can create artwork, graphics, logos, and illustrations
- Video tools like Sora are beginning to generate short video clips from written prompts
- Audio tools like ElevenLabs can generate realistic voiceovers and music

These are increasingly useful for small business owners, content creators, teachers, and anyone who needs visual or audio material without a big budget.

Productivity and Writing Assistance

These tools help you work faster and communicate better—drafting emails, organizing schedules, summarizing documents, improving your writing, and automating repetitive tasks.

- **Grammarly**—checks grammar, tone, and clarity in real time

- **Microsoft Copilot**—built into Word, Excel, and Outlook to help you write, analyze, and organize
- **Google Duet AI**—the same idea, built into Google Docs, Sheets, and Gmail

Code Generators and Developer Tools

Tools like GitHub Copilot help people write and fix computer code. If you are not a developer, you may not use these directly—but they are quietly making the software and apps you use every day better and faster.

Business Automation and Analysis

These tools connect your apps together, automate repetitive business tasks, and turn raw data into useful insights.

- **Zapier**—connects your tools and automates workflows without any technical knowledge requirement
- **HubSpot AI**—helps small businesses manage customers, marketing, and communications
- **Tableau and Power BI**—turn business numbers into visual charts and plain-language summaries

How to Choose the Right Tool

With so many options, the most practical approach is this:

- **Start with your need, not the tool.** Ask yourself: What problem am I trying to solve? What task am I spending too much time on? What do I wish I had help with? Find the tool built for that specific purpose.

- **Always try the free version first.** Nearly every major AI tool offers a free tier. Use it. Learn it. Decide whether it genuinely helps you before spending a single dollar. When the free version no longer meets your needs—when you find yourself hitting its limits regularly—that is the right time to consider a paid plan.

- **Learn one tool deeply before adding another.** It is tempting to try everything. Resist that temptation. One tool used well will do more for your life and business than five tools used poorly.

The Bottom Line

AI tools are everywhere—and that is a good thing. It means there is almost certainly a tool designed specifically for what you need, often available for free, and accessible without any technical background.

The goal is not to collect tools. The goal is to find the right ones, learn them well, and put them to work for your life, your family, and your legacy.

SAMPLE AI PROMPT

Sample Prompt You Can Try-Finding the Right AI Tool

I run a small flower shop with 2 employees. I want to use AI to save time, but I don't know where to start. What are the 3 most impactful AI tools I should try first, and exactly how would I use each one in my flower shop?

Sample AI Response

Wonderful—and great news: AI can make a meaningful difference for a small flower shop, even on a tight budget. Here are the three tools I would recommend starting with:

1. ChatGPT (for writing and planning)

Right now, your biggest time drain might be writing—social media captions, responses to customer inquiries, seasonal promotion announcements, or thank-you notes for wedding clients. ChatGPT can do all of this in seconds. Tell it: "Write an Instagram caption for our Mother's Day bouquet promotion, warm and elegant tone, under 100 words." You will have a polished draft instantly.

2. Canva (with its AI Magic Write feature)

Canva is a design tool that lets you create beautiful promotional graphics—even with no design experience. Its AI feature can generate layouts, suggest color combinations, and even write short text for you. Use it for Instagram posts, flyers, and seasonal promotions. The free version is excellent for most small shops.

3. Google Workspace with Gemini AI (for customer emails and scheduling)

If you use Gmail, Google's built-in AI assistant can help you draft professional replies to customer inquiries, summarize long email threads, and even help you compose quotes for event orders. It is built right into the tools you may already use daily.

Start with just one this week. Even fifteen minutes with ChatGPT can save hours of writing time every month.

Replace 'flower shop' with your own business or situation to get a personalized recommendation.

Chapter 8
AI Tools for Personal Use

Enhance, Don't Replace

Before we dive into the list, one word matters more than any other in this chapter: *enhance.*

AI does not do your living for you. It does not know your story, your voice, your history, or your heart. What it does is take what you bring and make it better, clearer, stronger, and more polished. That distinction is important—and it will save you a lot of frustration.

Here is a perfect example. Some people hear about AI and immediately think, "I'll just ask it to write my biography." They type "write my bio" and are disappointed when AI produces something generic and hollow. That is because AI does not know you. It cannot. What AI *can* do is take the biography you draft—in your own words, with your own details and your own story—

and enhance it. Polish the language. Strengthen the structure. Make it ready for a job application, a website, a speaking profile, or a book.

You bring the story. AI helps you tell it better.

That principle applies to everything in this chapter.

What AI Can Help You Create

The following are tasks AI can assist you with for personal and professional purposes. For each one, remember: your input, your details, and your direction always come first.

1. **Your Biography or Personal Profile**—Draft your own bio in your own words—your background, your accomplishments, your mission. Then ask AI to enhance it for a specific purpose: employment, a speaking engagement, a book jacket, or a social media profile. The result will sound like you—only sharper.
2. **A Curriculum or Course Outline**—Whether you are a teacher, a trainer, a minister, or a parent who wants to organize lessons for your children, AI

can help you build a structured curriculum from your ideas. Give it your topic, your audience, and your goals—and it will help you create a logical, professional outline.

3. **A Press Release**—Have an event, a book launch, a business opening, or an announcement to make? AI can help you write a professional press release that gets attention. Provide the who, what, when, where, and why—and AI will format it properly and make it ready to send.
4. **Articles and Blog Posts**—Whether you are sharing knowledge, building your brand, or establishing yourself as a voice in your community, AI can help you draft articles and blog posts. Give it your main points and your perspective—and it will help you turn your ideas into a polished, readable piece.
5. **Social Media Content**—Consistent social media presence matters—for personal brands, small businesses, ministries, and community organizations alike. AI can help you create captions, posts, content calendars, and engagement copy across platforms like Facebook, Instagram, and LinkedIn.

6. **Advertisements**—Need to promote a service, an event, or a product? AI can help you write ad copy that is clear, compelling, and targeted to the right audience. Tell it what you are selling, who you are selling to, and what action you want people to take—and it will draft the message.

7. **Manuscripts**—If you have a book inside you—a memoir, a faith story, a business book, a novel—AI can help you organize your thoughts, develop your chapters, and refine your language. You write. AI helps you shape and strengthen what you have written.

8. **Book Outlines**—Not sure how to structure your book? Give AI your subject, your audience, and your main message—and ask it to create a chapter-by-chapter outline. It is an excellent starting point that you can then customize with your own content and voice.

9. **Book Title and Subtitle**—Coming up with the right title is one of the hardest parts of writing a book. Give AI your topic, your audience, and the feeling you want your book to convey—and ask it to generate title and subtitle options. You may be surprised at what it produces.

10. **Math Problems**—AI can solve math problems step by step, showing every calculation and explaining every move in plain language. This is invaluable for students, for parents helping with homework, and for adults managing budgets, calculating interest, or working through financial decisions.
11. **Essays**—Students and adult learners can use AI to organize their thoughts, strengthen their arguments, improve their structure, and polish their writing. The key is to write your own ideas first—then use AI to help you express them more clearly and effectively.
12. **Meal Planning**—Tell AI your dietary needs, your budget, your health conditions, and your food preferences—and ask it to create a weekly meal plan with recipes and a shopping list. It is practical, personalized, and saves hours of planning time.
13. **Website Content**—Whether you need an About page, a services description, a mission statement, or product descriptions, AI can help you write clear, professional website content that represents you well. Provide your details and your tone—and AI will help you find the right words.

14. **And Much More**—The list above is just the beginning. AI can also help you write thank-you letters, prepare for job interviews, plan events, draft business proposals, create presentations, research topics, and organize your ideas in virtually any area of life.

One Final Reminder

AI is a remarkably capable assistant—but the best results always start with you. The more clearly you communicate what you need, the better AI performs. Give it your details, your voice, your goals, and your context. Let it do the heavy lifting on structure and polish. And always review what it produces before you use it.

You are not outsourcing your thinking. You are amplifying it.

SAMPLE AI PROMPT

Sample Prompt You Can Try—Personal Trip Planning with AI

Help me plan a 5-day trip for two adults (both in our 60s, one with a bad knee) to Washington D.C. We love history, good food, and art museums. We prefer to walk slowly and avoid very crowded places. We have a moderate budget. Give me a day-by-day itinerary.

Sample AI Response

What a wonderful trip to plan! Washington D.C. is perfect for history lovers—and very manageable for someone with a knee concern if you pace yourselves wisely. Here is a gentle, enriching five-day itinerary:

Day 1—Arrival and the National Mall (Easy Pace)

Arrive and settle in. In the afternoon, take a slow stroll along the National Mall. The Lincoln Memorial and the Reflecting Pool are flat, paved, and stunning. Have dinner at Old Ebbitt Grill, a Washington institution just blocks from the White House.

Day 2—Smithsonian Day

The Smithsonian museums are free and spread across the Mall. Pick two: the National Museum of American History and the National Portrait Gallery are favorites. Both have excellent benches for resting and are not too crowded on weekday mornings.

Day 3—Art and Culture

The National Gallery of Art is world-class and has motorized carts available if needed. Spend a leisurely morning there. Afternoon: the quiet, beautiful U.S. Botanic Garden nearby offers a lovely, seat-friendly rest.

Day 4—Georgetown and a Scenic Drive

Georgetown's waterfront is beautiful and walkable along flat brick paths. Consider hiring a car for a drive past the White House, Capitol, and Embassy Row—all the sights with minimal walking.

Day 5—Arlington and Departure

The Arlington National Cemetery is profoundly moving. Trams are available throughout the grounds—no extensive walking required.

Pro tip: Book a hotel near the Metro with easy elevator access, and always call ahead to confirm accessibility features.

Providing your specific constraints—age, mobility, budget, interests—leads to genuinely personalized results.

Chapter 9
AI Tools for Business—Work Smarter, Grow Faster

The Playing Field Has Changed

For decades, small business owners and entrepreneurs operated at a disadvantage. Large corporations had teams of designers, writers, marketers, lawyers, and analysts. Small businesses had themselves—and whatever they could afford to hire.

AI has changed that equation entirely.

Today, a solo entrepreneur, a small business owner, a ministry leader, or a community organization can access the same quality of tools that once required an entire department. The cost of entry is low. The potential return is enormous. And the time savings alone can transform how you run your business.

When you put AI to work in your business, you are not just saving time. You are reclaiming your focus—shifting your energy away from working *in* your business and toward building *your* business.

What AI Can Do for Your Business

Here is a practical look at what AI tools can help you create, automate, and accomplish:

1. **Automate Repetitive Tasks** Every business has tasks that happen over and over—scheduling appointments, sending follow-up emails, organizing files, responding to common customer questions. AI can handle these automatically, around the clock, without fatigue and without error. Automating repetitive work is like hiring a virtual assistant who never calls in sick and never asks for overtime.

   The time you reclaim goes directly toward what matters most: building relationships, growing revenue, and strengthening your vision.

2. **Build Your Own Website** AI-powered website builders can help you create a professional, polished website without hiring a developer or

learning any code. Describe your business, your services, and your style—and AI will help you build a site that represents you well.

3. **Design Your Company Logo** Tools like Canva AI and Adobe Firefly can generate professional logo options based on your business name, industry, and preferred style. What once cost hundreds of dollars in design fees can now be accomplished in minutes.

4. **Create Business Cards and Marketing Materials** AI design tools can produce business cards, flyers, brochures, and promotional materials that look professionally designed—customized to your brand colors, fonts, and message.

5. **Write and Design a Newsletter** Staying in front of your customers consistently is one of the most powerful things a small business can do. AI can help you write, design, and schedule a regular newsletter that keeps your audience engaged and your brand visible.

6. **Develop a Professional Photo or Glamour Shot** AI image tools can enhance your existing photos to look polished and professional—suitable for your website, your LinkedIn profile, your press kit, or your book jacket. No expensive photo shoot required.

7. **Write Press Releases** When you have news to share—a business launch, a new service, an award, a community event—AI can help you write a properly formatted, professional press release ready to send to media outlets and post online.

8. **Create Articles and Blog Content** Consistent content builds credibility and drives traffic to your business. AI can help you generate articles, blog posts, and thought leadership content that positions you as an authority in your field. Provide your ideas and your expertise—AI helps you express them clearly and consistently.

9. **Write Ad Copy** Whether you are running ads on Facebook, Instagram, Google, or in print, AI can help you write compelling, targeted advertising

copy that speaks directly to your ideal customer and moves them to action.

10. **Manage Social Media Content** AI can help you plan, write, and schedule social media posts across multiple platforms—maintaining a consistent presence without consuming your entire day. Give it your message, your audience, and your tone—and let it help you stay visible.

11. **Produce Promotional Videos** AI video tools can help you create short promotional videos for your business, your services, or your events—without a film crew or a big budget. These are increasingly powerful for social media marketing and customer engagement.

12. **Launch a Podcast** AI tools can help you outline episodes, write scripts, edit audio, generate transcripts, and even create promotional materials for a podcast. If you have knowledge and a message to share, AI lowers every barrier to getting it out into the world.

13. **Develop a Business Plan** A strong business plan is essential for securing funding, attracting

partners, and staying focused on your goals. AI can help you build a professional, thorough business plan—covering your mission, market analysis, financial projections, and growth strategy—starting from your own ideas and vision.

14. **Write Curriculums and Training Materials** If your business includes coaching, teaching, training, or ministry, AI can help you develop structured curriculums, course outlines, workshop materials, and training guides that are organized, professional, and effective.

15. **Apply for Grants** Grant writing is one of the most valuable—and most time-consuming—tasks for nonprofits, small businesses, and community organizations. AI can help you research available grants, understand requirements, and draft compelling grant applications that represent your mission powerfully.

16. **Write Business Proposals** Whether you are pitching a new client, pursuing a contract, or seeking a partnership, AI can help you write

a professional business proposal that clearly communicates your value, your approach, and your terms.

17. **Draft Contracts and Business Agreements** AI can help you create first drafts of standard business contracts, service agreements, partnership documents, and other legal templates. Always have a licensed attorney review any contract before signing or presenting it—but AI can help you get a solid draft on the table quickly and affordably.

18. **And Much More** Customer service chatbots. Inventory management. Financial forecasting. Employee onboarding materials. Translation for global markets. The applications are virtually limitless—and growing every day.

The Competitive Advantage You Cannot Afford to Ignore

AI tools also allow you to collaborate in real time with customers in different languages—giving even the smallest business access to a global market. A home-based business in Atlanta can now serve clients in London, Lagos, or Lima with the same professionalism as a multinational corporation.

The businesses that embrace AI now will have a significant and compounding advantage over those that wait. Every hour saved on repetitive tasks is an hour invested in growth. Every dollar saved on design, writing, and administration is a dollar reinvested in your vision.

You do not need a large team. You do not need a large budget. You need the willingness to learn the tools and the discipline to use them consistently.

That is the new competitive advantage. And it is available to you right now.

SAMPLE AI PROMPT

Sample Prompt You Can Try—Generating Real Business Content

Write a professional email newsletter for a family-owned Italian restaurant called 'Bella Nonna' to send to loyal customers announcing a new Sunday brunch menu launching next month. Include a special offer for email subscribers: 15% off their first brunch visit. Tone: warm, inviting, like a letter from the family. Under 250 words.

Sample AI Response

Subject: Sunday Brunch Is Coming—And It's Made With Love

Dear Friends of Bella Nonna,

Something wonderful is happening in our kitchen.

For years, you have joined us for dinner—for birthdays, anniversaries, quiet Tuesday evenings, and the moments in between that deserve a good meal and a glass of something beautiful. You have become part of our family, and we do not take that lightly.

That is why we are so excited to tell you: starting next month, Bella Nonna is opening on Sunday mornings for brunch.

Not just any brunch. Our brunch.

Think: pillowy ricotta pancakes, a frittata that changes with the season, freshly baked cornetti, and Nonna's tomato sauce—yes, that one—spooned over slow-baked eggs in the way she used to make on Sunday mornings back in Puglia.

We will open at 10:00 a.m. and close when the afternoon takes us there.

And because you have been with us from the beginning, we want you to be first through the door. As a loyal email subscriber, you will receive 15% off your first Sunday brunch visit. Simply mention this email when you arrive.

Reservations open next week. We expect tables to fill quickly.

With gratitude and garlic,

The Bella Nonna Family

This type of content would typically take 30-60 minutes to write. AI produced a polished draft in seconds.

Chapter 10
Samples of Prompts for Personal Use

See It in Action

Reading about AI is one thing. Seeing it work is another. This chapter shows you real prompts—written the way you would actually type them—alongside the real responses AI provided. These are not made-up examples. They are the kind of practical, everyday situations that real people face, and they show you exactly what AI can do when you give it clear, specific information.

Study the prompts carefully. Notice how much detail is included. That detail is what produces results this useful.

EXAMPLE 1: Meal Planning for a Senior with Medical Dietary Needs

The Prompt: *"I need a 7-day meal plan for my mother. She is seventy years old and she is on an 1,800-calorie diabetic low-sodium diet."*

Why This Prompt Works: It provides the person's age, the calorie target, and two specific dietary requirements. AI has everything it needs to produce something genuinely useful—not generic.

The Response:

This plan is built around approximately 1,800 calories per day, with balanced carbohydrates, heart-healthy fats, lean protein, and no added salt. Use herbs, lemon, garlic, onion, pepper, and salt-free seasoning blends in place of sodium.

Daily Targets:

- Carbohydrates: 45–60 grams per meal, 15–20 grams per snack
- Sodium: 1,500 mg or less per day

- Protein: included with every meal
- Cooking methods: baked, grilled, steamed, or sautéed with olive oil spray

Day 1

Breakfast: Oatmeal (½ cup cooked), blueberries (½ cup), boiled egg, unsweetened tea or coffee Snack: Small apple + 1 tablespoon peanut butter (no added salt) Lunch: Grilled chicken breast, quinoa (½ cup), steamed green beans Snack: Plain Greek yogurt (½ cup) with cinnamon Dinner: Baked salmon, brown rice (½ cup), steamed broccoli, side salad with olive oil and vinegar

Day 2

Breakfast: Scrambled eggs (1 whole egg + 2 egg whites), whole-grain toast (1 slice), small orange Snack: Handful of unsalted almonds Lunch: Turkey lettuce wrap, small baked sweet potato Snack: Low-sodium cottage cheese with sliced cucumber Dinner: Baked skinless chicken thigh, cauliflower mash, steamed carrots

Day 3

Breakfast: Greek yogurt parfait—plain yogurt, strawberries, 2 tablespoons low-sugar granola Snack:

Small pear Lunch: Tuna salad (olive oil and lemon, no mayonnaise), 5–6 whole-grain crackers Snack: Celery sticks with low-sodium hummus Dinner: Turkey meatloaf (no added salt), mashed sweet potatoes, green beans

Day 4

Breakfast: Oatmeal with chopped walnuts, cinnamon, and half a banana Snack: Hard-boiled egg Lunch: Homemade lentil soup (no added salt), side salad Snack: Small handful of grapes Dinner: Baked cod, wild rice (½ cup), sautéed spinach in olive oil

Day 5

Breakfast: Veggie omelet with onions, peppers, and spinach; whole-grain toast (1 slice) Snack: Apple slices with almond butter Lunch: Grilled chicken salad with olive oil and vinegar dressing Snack: Plain yogurt or no-sugar protein shake Dinner: Lean beef stir-fry, brown rice (½ cup), mixed vegetables

Day 6

Breakfast: Smoothie—unsweetened almond milk, mixed berries, diabetic-friendly protein powder Snack: Low-sodium cheese stick Lunch: Baked turkey burger (no bun), roasted vegetables Snack: Handful of unsalted nuts Dinner: Grilled salmon, sweet potato, asparagus

Day 7

Breakfast: Whole-grain waffle (1), small amount of sugar-free syrup, low-sodium turkey sausage Snack: Orange or kiwi Lunch: Homemade chicken vegetable soup, side salad Snack: Cottage cheese with berries Dinner: Roast chicken, brown rice (½ cup), steamed cabbage

What This Example Teaches: AI can create a detailed, medically informed meal plan in seconds—one that would take hours to research and compile manually. Always share results like this with your mother's doctor or dietitian to confirm it aligns with her specific medical needs and medications.

EXAMPLE 2:
Understanding a Medication's Side Effects

The Prompt: *"My 85-year-old father was hospitalized for sepsis. Before being hospitalized, he had a little dementia. He has prostate cancer and was prescribed Bicalutamide. Since taking this medication, he is confused, hallucinating, delusional, and disoriented. What are the side effects of this medication?"*

Why This Prompt Works: The family member provided the patient's age, existing conditions, current situation, the specific medication name, and the exact symptoms of concern. AI can now give a focused, relevant response—not a generic drug summary.

The Response:
Bicalutamide is a non-steroidal anti-androgen commonly used to treat prostate cancer, often alongside other therapies. Like many hormone-affecting medications, it carries a range of side effects—some common and manageable, some less common but important, and some that require immediate medical attention.

Common Side Effects:

- Hot flashes and sweating
- Breast tenderness or enlargement
- Decreased sex drive
- Fatigue or weakness
- Weight gain
- Nausea, diarrhea, or constipation
- Dizziness
- Back or pelvic pain

Less Common but Important Side Effects:

- Liver problems—elevated enzymes, yellowing of skin or eyes, dark urine
- Anemia
- Shortness of breath
- Depression or mood changes
- Skin rash or itching
- Loss of appetite

Rare but Serious—Seek Medical Care Immediately:

- Severe confusion or disorientation
- Hallucinations or delusions
- Significant worsening of existing cognitive conditions such as dementia

What This Example Teaches: The symptoms this father was experiencing—confusion, hallucinations, delusions, and disorientation—fall into the serious category, especially in a patient who already had dementia before starting the medication. This response gave the family the language and the knowledge to go back to the medical team and ask specific, informed questions. That is exactly what AI is for. It does not replace the doctor. It helps you talk to the doctor with confidence.

EXAMPLE 3:
Writing a Professional Letter for a Personal Emergency

The Prompt: *"I need a letter written for my supervisor notifying her that I have a medical emergency in my family. I must leave the city immediately. This is beyond my control. My absence will extend beyond six weeks, and I would appreciate any consideration for my returning to my position. I really need my job. I am head of household."*

Why This Prompt Works: The person explained the situation clearly, stated the length of absence, expressed the need to protect their job, and shared personal context. AI used all of that to write a letter that is professional, dignified, and emotionally appropriate.

The Response:
Subject: Emergency Leave Request Due to Family Medical Crisis
Dear [Supervisor's Name],
I am writing to formally notify you that I am facing a

serious and unforeseen medical emergency involving an immediate family member. Due to the severity of the situation, I must leave the city immediately to provide care and support. This circumstance is entirely beyond my control.

At this time, I anticipate that my absence may extend beyond six weeks, though I will remain in communication as circumstances evolve. I am committed to providing any documentation required and to following all appropriate procedures related to leave.

I respectfully request your consideration and understanding regarding my ability to return to my position once this emergency has stabilized. My employment is extremely important to me, as I am the head of household, and I am fully committed to resuming my duties as soon as I am able.

Thank you for your time, compassion, and any consideration you are able to extend during this very difficult period. I look forward to discussing next steps when possible.

Sincerely, [Your Full Name] [Your Job Title] [Your Department] [Your Phone Number]

What This Example Teaches: In one of the most stressful moments of a person's life—a family emergency, a job at risk, an urgent departure—AI produced a professional, dignified letter in seconds. The person did not need to know the right words. They just needed to explain their situation clearly. AI did the rest. Replace the bracketed fields with your information, review it once, and send.

The Lesson Behind All Three Examples

Look at what these three prompts have in common. Each one:

- Provided specific details about the person and the situation
- Had a clear goal in mind
- Trusted AI with enough context to produce something genuinely useful

That is the formula. The more you put in, the more you get out. And what you get out can save you time, protect your family, and help you navigate life's most challenging moments with greater clarity and confidence.

SAMPLE AI PROMPT

Sample Prompt You Can Try—Elder Care Guidance

My mother is 80 years old and was just diagnosed with mild cognitive decline. She still lives alone and wants to stay independent as long as possible. As her daughter, what are 5 practical, compassionate steps I can take right now to support her independence and safety—without making her feel like a burden?

Sample AI Response

This is one of the most loving things a daughter can do—and your instinct to support without diminishing her dignity is exactly right. Here are five steps to consider:

1. Have an honest, gentle conversation with her—not about limitations, but about planning.

Frame it as something you are doing together: "Mom, I want us both to feel prepared and confident. Can we spend some time thinking through what would help you feel most comfortable at home?" This invites her as a partner, not a patient.

2. Simplify her environment gradually and unobtrusively.

Remove rugs that could cause falls. Ensure pathways are well-lit. Consider a medical alert button she wears—many now look like a simple watch. Make these changes quietly and frame them positively: "I got you this watch—I love the look of it."

3. Create a gentle daily rhythm and write it down together.

A simple, consistent routine reduces the cognitive load of decisions. A whiteboard in the kitchen with the day's plan ("Tuesday: Walk, Lunch, Call with Sarah") can be enormously reassuring without feeling clinical.

4. Build in regular, natural contact—without it feeling like checking up.

A daily phone call that is genuinely about connection ("I was thinking about you and that pie you used to make") keeps you informed while preserving her sense of independence and dignity.

5. Connect with her doctor now, while things are stable.

Ask her permission to be included in medical conversations. Establish yourself as a point of contact before a crisis requires it. This is not taking over—it is being prepared together.

You are doing this beautifully already simply by asking the right questions.

AI can help you prepare for sensitive conversations, explore options, and feel less alone in difficult situations.

Chapter 11
Samples of Prompts for Business

Real Prompts. Real Results. Real Business.

Chapter 10 showed you what AI can do for your personal life. This chapter shows you what it can do for your business. The same principle applies: the more specific and detailed your prompt, the more powerful and useful the response.

The examples in this chapter cover four very different business needs—a complete business plan, a book outline and title, a rental advertisement, and multilingual marketing. Each one demonstrates a different dimension of what AI can produce for you quickly, professionally, and at no cost beyond your time. Study the prompts. Notice the detail. Then imagine how you could apply the same approach to your own business.

EXAMPLE 1:
Building a Complete Business Plan

The Prompt: *"What are the elements of a business plan for a community bookstore?"*

Why This Prompt Works: It is specific about the type of business—not just any business, but a community bookstore—which allows AI to tailor every section to that mission, audience, and funding environment.

The Response:

Business Plan: Community Bookstore

1. Executive Summary *Write this last—place it first.* This is your "why it matters" section. Include:

- Business name and location
- Mission statement: community literacy, local culture, support for local authors, safe gathering space
- Target community served
- Products and services offered: books, events, café, workshops, and more

- Legal structure: LLC, nonprofit, or hybrid
- Startup costs and funding needs
- Three- to five-year vision

2. Mission, Vision, and Community Impact *Critical for grants and partnerships.*

- Mission statement—why you exist
- Vision statement—what success looks like
- Community impact goals: literacy improvement, support for local and independent authors, youth engagement, safe gathering space, economic revitalization

3. Business Description Explain exactly what kind of bookstore this is:

- New or existing business
- Brick-and-mortar, mobile, or hybrid model
- For-profit, nonprofit, or fiscally sponsored
- Cultural focus: Black literature, children's books, faith, social justice, local history
- How you differ from Amazon and big-box retailers

4. Market Analysis Show that you understand your environment:

- Community demographics and literacy levels
- Local schools, churches, and community organizations
- Competition from other bookstores and online retailers
- Gaps you will fill and customer pain points you will address

5. Products and Services *Beyond selling books.*

Core offerings:

- New and used books
- Children's books
- Local and independent author titles
- Educational materials

Community revenue streams:

- Author events and book signings
- Writing workshops
- Reading circles and book clubs
- Youth programs
- Open mic nights
- Café or snack bar
- Branded merchandise: totes, journals, T-shirts

6. Programs and Events *This is what turns a bookstore into a community anchor.*

- Weekly story time
- School partnerships
- Literacy nights
- Senior reading clubs
- Faith and family discussions
- Small business pop-ups
- Open mic and spoken word events

7. Marketing and Outreach Strategy How people will find you:

- Branding and messaging
- Social media presence and email list
- Partnerships with schools, churches, and nonprofits
- Local radio and talk shows
- Author networks and community ambassadors
- Website and online store

8. Operations Plan How the store will run daily:

- Location and hours of operation
- Staffing: paid employees versus volunteers

- Inventory management system
- Point-of-sale system
- Supplier relationships
- Security and insurance
- Accessibility: ADA compliance, seniors, children

9. Management and Organization Who is in charge:

- Founder and owner biography
- Management team structure
- Board of advisors or directors (if nonprofit)
- Roles, responsibilities, and volunteer coordination

10. Legal Structure and Compliance Important for protection and funding eligibility:

- LLC versus nonprofit versus hybrid structure
- Business licenses and seller's permit
- Zoning requirements
- Insurance coverage
- Sales tax compliance

11. Financial Plan *This is where many plans fail—do not skip it.*

- Startup costs itemized
- Monthly operating expenses
- Pricing strategy
- Sales projections for 12 to 36 months
- Break-even analysis
- Cash flow projections
- Funding sources: grants, loans, donors, investors

12. Funding Strategy *Especially important for community bookstores:*

- Literary, arts, and community development grants
- Corporate sponsorships
- Individual donations
- Crowdfunding campaigns
- City and county programs
- Partnerships with foundations

13. Risk Analysis and Sustainability Show that you are realistic:

- Economic downturns and their impact
- Online retail competition
- Rising rent and overhead costs

- Staffing challenges
- Contingency plans
- Diversified income streams

14. Appendix Supporting documents:

- Floor plan
- Letters of support from community organizations
- Community needs data
- Founder and team resumes
- Sample programming calendar

What This Example Teaches: A complete, professional business plan outline—tailored specifically to a community bookstore, including grant strategy, cultural programming, and community impact—produced in one prompt. This is a framework you can build on immediately. Fill in your specific details, customize the sections to your vision, and you have the foundation of a fundable, credible business plan.

EXAMPLE 2:
Creating a Book Outline, Title, and Subtitle

The Prompt: *"I need a book outline, title, and subtitle for a book teaching parents how to maximize the talents and abilities of a challenged child—especially when the child wants to mimic what other children are doing."*

Why This Prompt Works: It identifies the audience (parents), the subject (challenged children), and a specific emotional situation (the child wanting to fit in with peers). AI uses all of that context to produce something genuinely tailored and useful.

The Response:

Book Title and Subtitle Options

Option 1—Warm and Empowering: **Different, Not Defeated** *How Parents Can Unlock the Gifts, Build Confidence, and Safely Guide a Challenged Child Toward Independence*

Option 2—Faith and Practical: **God Gave Them a Gift** *Raising a Challenged Child to Discover Purpose, Ability, and Courage in a World That Moves Fast*

Option 3—Straight Talk for Parents: **Let Me Try** *A Parent's Guide to Helping a Challenged Child Grow, Learn, and Safely Find Their Own Way*

Option 4—Independence Focused: **Teaching Wings Without Letting Go** *How to Protect, Prepare, and Empower a Challenged Child Who Wants to Do What Other Children Do*

Option 5—Legacy and Long-Term: **Beyond the Label** *Preparing a Challenged Child for Confidence, Independence, and a Life of Purpose*

Full Book Outline

Introduction: I See Your Child—and I See You

- The quiet fears parents do not say out loud
- Loving your child without shrinking their world
- Why "challenged" does not mean "limited"

Part I: Seeing the Child Beyond the Challenge

Chapter 1: Your Child Is Not Broken

- Separating diagnosis from destiny
- Letting go of shame, guilt, and comparison
- Understanding that ability shows up differently

Chapter 2: The Gift Hidden Inside the Struggle

- Every child has a strength—even if it is not academic
- Discovering talents through observation, not pressure
- When creativity, empathy, memory, or persistence is the gift

Part II: When Your Child Wants to Do What Other Children Do

Chapter 3: "Why Can't I Do What They Do?"

- The emotional weight of comparison
- Why mimicking peers is natural and healthy
- Helping your child feel included without setting them up for danger

Chapter 4: Independence Without Injury

- Teaching "how" instead of saying "no"
- Breaking activities into safe, manageable steps

- Knowing when supervision builds confidence instead of fear

Chapter 5: Teaching Safety Without Killing Confidence

- How to talk about danger without creating panic
- Repetition, routines, and clear boundaries
- Teaching your child to ask for help without embarrassment

Part III: Building Skills for Real Life

Chapter 6: Teaching Life Skills Early

- Self-care, basic money skills, and communication
- Age-appropriate responsibility
- Why independence starts at home

Chapter 7: Social Skills, Peer Pressure, and Being Left Out

- Helping your child navigate friendships
- Teaching discernment—not fear
- Preparing them for unkind words and moments of exclusion

Part IV: Partnering with Schools, Family, and Community

Chapter 8: Advocating Without Apologizing

- Managing exhaustion and burnout
- Finding joy in the journey

Conclusion: Let Them Try—With Wisdom, Love, and Guardrails

- Why dignity matters more than safety alone
- Why confidence changes everything
- How your belief becomes their courage

Bonus Sections:

- Parent Reflection Questions
- Safety Checklists
- "Let Me Try" Conversation Scripts
- Talent Discovery Worksheets
- A Letter to Your Child

What This Example Teaches: One prompt produced five title options, a full multi-part outline, and a complete chapter-by-chapter structure—including bonus content. What might have taken days of planning was delivered in seconds. Take the title that resonates with your voice, customize the chapters with your own stories and expertise, and you have a publishable book framework ready to develop.

EXAMPLE 3:
Writing a Rental Advertisement

The Prompt: *"I need a rental ad for my three-bedroom home in Los Angeles, California. It has three bedrooms, 2½ baths, an island kitchen, wet bar in the den, large lot, double garage, new roof, and is near schools, markets, and major freeways. It is zoned for an additional dwelling unit. Mixed neighborhood with friendly neighbors and an active block club. Ready for move-in."*

Why This Prompt Works: Every key feature of the property is listed, along with the neighborhood's social character. AI uses all of it to write copy that sells the home's lifestyle, not just its square footage.

The Response:

FOR RENT—SPACIOUS 3-BEDROOM HOME IN LOS ANGELES *Now available and ready for immediate move-in!*

This well-maintained 3-bedroom, 2½-bath single-family home offers comfort, space, and future potential in an established Los Angeles community.

Property Features:

- 3 bedrooms | 2½ bathrooms
- Large kitchen with center island
- Wet bar in the den—perfect for entertaining
- Double garage
- New roof
- Large lot zoned for an additional dwelling unit—excellent income or extended-family potential

Location Highlights:

- Conveniently located near schools, grocery markets, and major freeways
- Mixed neighborhood with friendly neighbors
- Active and supportive block club

Location: Los Angeles, CA | Available: Immediate move-in Contact: [Your Name / Phone / Email] Serious inquiries only.

EXAMPLE 4:
Translating the Same Ad into Spanish

The Prompt: *"Write the same ad in Spanish also."*

Why This Prompt Works: This is one of the most powerful demonstrations of AI's business value. One follow-up sentence—no additional detail required—produced a complete, professional translation. This is what global market access looks like for a small business owner.

The Response:

SE RENTA—CASA ESPACIOSA DE 3 RECÁMARAS EN LOS ÁNGELES *¡Disponible ahora y lista para mudarse de inmediato!*
Esta casa unifamiliar bien mantenida cuenta con 3 recámaras y 2½ baños, ofreciendo comodidad, amplitud y potencial a futuro en una comunidad establecida de Los Ángeles.

Características de la Propiedad:

- 3 recámaras | 2½ baños
- Cocina amplia con isla central
- Bar húmedo en el área familiar—ideal para convivencias
- Garaje doble
- Techo nuevo
- Terreno grande zonificado para vivienda adicional—excelente potencial de ingresos o para familia extendida

Ventajas de la Ubicación:

- Cerca de escuelas, mercados y autopistas principales
- Vecindario mixto con vecinos amigables
- Club de cuadra activo y solidario

Ubicación: Los Ángeles, CA | Disponibilidad: Inmediata
Contacto: [Su nombre / Teléfono / Correo electrónico]
Solo interesados serios.

What Examples 3 and 4 Teach Together: In minutes, AI produced a professional rental listing ready to post on Zillow, Craigslist, Facebook Marketplace, or any rental platform—and then translated the entire ad into Spanish with a single follow-up prompt. For property owners, landlords, and real estate investors operating in diverse communities, this capability alone is worth the price of learning AI. Which, as you know by now, is free.

The Pattern Behind Every Successful Business Prompt

Look at what every example in this chapter has in common:

- The prompt was **specific** about the type of business or need
- It included **relevant details**—location, audience, features, purpose
- It had a **clear goal**: a plan, an outline, an ad, a translation
- It treated AI as a **starting point** to be customized, not a finished product to be used without review

That pattern works across every industry, every business size, and every budget. Master it, and AI becomes one of the most powerful business tools you have ever had access to—available to you right now, at no cost, on any device with an internet connection.

Chapter 12
Conclusion

The New Literacy

We began this journey with a simple but powerful premise: AI matters to everyone.

Not just to engineers. Not just to corporations. Not just to the young, the wealthy, or the technically gifted. AI matters to the senior who wants to protect what she has built over a lifetime. It matters to the parent who wants to give their children a fighting chance. It matters to the student who has been told, directly or indirectly, that certain doors are not open to her. It matters to the small business owner, the community leader, the minister, the grandmother, and the grandchild sitting beside her.

Artificial Intelligence is the new literacy.

Think about that carefully. Imagine how difficult life is for someone who cannot read or write English and is trying to survive in America—navigating forms, contracts, job applications, medical instructions, and financial documents without the ability to understand them fully. Now imagine that same disadvantage, multiplied, for people in the very near future who are AI-illiterate. Within the next five years, not understanding AI will put people at a serious disadvantage—economically, professionally, and personally.

This is not a prediction meant to frighten you. It is a warning meant to move you.

AI Readiness Is Not a Fad

AI is not a trend that will pass. It is not a phase that will fade. It is not optional for those who want to remain economically competitive, professionally relevant, or personally empowered.

Artificial Intelligence will reshape how we live, work, earn, and survive. It will create opportunities for some and real hardship for others. The difference between those two outcomes will not be intelligence, education level, or age. The difference will be preparation.

Those who understand AI will move ahead. Those who ignore it risk being left behind—not because they are incapable, but because they were not prepared. And that is a circumstance within your control, starting today.

What You Now Know

Over the course of this book, you have traveled through the full landscape of AI—from its origins in the 1950s to the tools available on your phone right now. You have learned the language: prompts, large language models, machine learning, hallucinations. You have seen real prompts produce real results—meal plans, medication research, business plans, book outlines, rental ads, emergency letters, and more.

You have learned the dos and don'ts. You have seen the difference between a vague prompt and a specific one. You have discovered that AI does not require a college degree, a technical background, or youth. It requires only one thing: the willingness to learn.

If an 80-year-old woman can work with AI the way a farmer works his mule—every day, steadily, purposefully, and without fear—so can you.

Your Responsibility Does Not End Here

Knowledge kept to yourself is only half as powerful as knowledge shared.

I urge you—with everything in me—to share this book with your family, your friends, and your loved ones. Pass it to the senior in your church who thinks technology is not for her. Give it to the young parent who is overwhelmed and does not know where to start. Put it in the hands of the student who has been made to feel that the future is not for people like him.

Do not keep this information to yourself.

In fact, I challenge you to become more than a reader. Become a messenger. Take responsibility for spreading this knowledge in your home, your church, your school, and your community. Host a conversation. Organize a workshop. Sit down with a neighbor and walk through a prompt together. Every person you bring along with you multiplies the impact of what you have learned.

Our joy, our peace, and our future prosperity are connected to the well-being of the people around us. When our community is informed, prepared, and empowered—we all rise together.

A Final Word

Legacy is not just what you leave. It is what you prepare the next generation to receive.

AI is a tool. Wisdom governs the tool. And wisdom, freely shared, becomes the foundation of a legacy that outlasts any single life.

You picked up this book because something in you refused to be left behind. That same spirit—that refusal, that hunger, that willingness—is all you have ever needed. It is all you need now.

Go learn. Go share. Go rise.

And take somebody with you.

Thank you for reading *AI Made Simple*.

May your journey with AI be curious, empowered, and deeply human.

Books by Dr. Rosie Milligan

Empowering Readers Through Knowledge, Entrepreneurship, and Personal Growth.

Business Books

- What You Need to Know Before Starting Your Business
- LLC & S Corp Nuggets Made Simple
- How to Close Your Business Legally and Gracefully Made Simple
- LLC & Corporation Essentials – Your Ultimate FAQ Guide for Entrepreneurs Made Simple

Finance Books

- Getting Out of Debt Made Simple
- Creating A Budget Made Simple
- Understanding Credit Made Simple

Self-Help Books

- Creating A New You Made Simple
- What To Do Before You Get Hitched
- Satisfying the Black Man Sexually Made Simple
- Satisfying the Black Woman Sexually Made Simple
- AI (Artificial Intelligence) Made Simple
- AI Made Simple for Parents
- Departing This Life Preparations

Books for Writers

- Nuts & Bolts For the New Author & New Publisher Made Simple
- ABC's On How To Prepare Your Manuscript for Editing, Formatting and Printing
- How to Write A Book Made Simple & Your Publishing Options
- The ABC's For Starting & Managing Your Own Publishing House Made Simple
- Developing A Marketing Plan For Your Books Made Simple

The Dr. Rosie Milligan Knowledge Series

This powerful collection of books by Dr. Rosie Milligan provides practical knowledge designed to help individuals build stronger lives, businesses, and communities.

Business Builder Series

Helping entrepreneurs start, manage, and protect their businesses.

Financial Freedom Series

Teaching readers how to manage money, eliminate debt, and build financial stability.

Life & Personal Growth Series

Helping individuals strengthen relationships, transform their lives, and embrace new knowledge.

Artificial Intelligence Series

Making AI simple for adults, parents, and the next generation.

Author & Publishing Series

Guiding writers on how to write, publish, and market their books successfully.

Dr. Rosie Milligan – Author, Publisher, Educator, and Founder of Black Writers On Tour.

www.ingramcontent.com/pod-product-compliance
Lightning Source LLC
LaVergne TN
LVHW101616210926
848480LV00009B/593
* 9 7 9 8 9 9 4 1 3 0 9 1 9 *